George W. Peckham, Elizabeth G. Peckham

Description of New or Little Known Spiders of the Family

Attidae

from various parts of the United States of North America

George W. Peckham, Elizabeth G. Peckham

Description of New or Little Known Spiders of the Family Attidae
from various parts of the United States of North America

ISBN/EAN: 9783337300326

Printed in Europe, USA, Canada, Australia, Japan

Cover: Foto ©berggeist007 / pixelio.de

More available books at **www.hansebooks.com**

DESCRIPTIONS OF NEW OR LITTLE KNOWN SPIDERS OF THE FAMILY ATTIDÆ, FROM VARIOUS PARTS OF THE UNITED STATES OF NORTH AMERICA.

WITH THREE PLATES.

BY G. W. AND E. G. PECKHAM.

In describing the following spiders we have used but three generic names; not because we consider all these species as properly belonging together, but inasmuch as we are now engaged in preparing a monograph of this family, we judge it best to defer any discussion of its classification until later. Meanwhile we shall hope to prepare the way for this discussion by shortly publishing a synopsis of the various genera thus far formed. Although we have many Attidæ from different parts of the world, we shall feel greatly obliged to any arachnologists who will aid us in our difficult undertaking by sending us collections of Attidæ. For all such collections we will gladly send, in return, spiders of the United States.

We are indebted to Mr. Rudolph Haessler for the drawings, which were made from nature.

MILWAUKEE, WIS., October 10th, 1883.

I. ATTUS PUTNAMII, new.

(Plate 1, Figures 1 to 1b.)

MALE Length, 9 mm.; width of abdomen, 2.6 mm.; length of ceph.th. 4.5 mm.; width of ceph.th., 3.3 mm.

The cephalothorax is high, highest and widest at the dorsal eyes, the thoracic part being longer than the cephalic. The color is reddish brown; there is a large white spot just behind the dorsal eyes, and a short white band on each side of the eye-region extending from the small median to the dorsal eye. Just back of the small median eye is a tuft of black hairs. There are thick, light brownish hairs above the anterior row of eyes.

The quadrangle of the eyes is wider than long, and much wider behind. The dorsal eye is about as large as the lateral eye, and is placed higher, so that a straight line from its lower border cuts the upper border of the lateral eye. The small median eye is in a line with the dorsal eye, but is further from the dorsal than from the lateral eye. The anterior row of eyes is very slightly curved, a straight line from the top of the middle eyes cutting the upper borders of the lateral eyes. The lateral are well separated from the middle eyes; the middle eyes touch each other. The middle are not quite twice as large as the lateral eyes. The clypeus is two-thirds as wide as the middle eyes, vertical, covered with short red and long white hairs.

The palpus is brownish, with black and white hairs. The tibia is very short, bearing no apophysis ; the tarsus is cut obliquely at the extremity ; the bulb is simple.

The falces are stout, vertical, slightly longer than the face ; they are covered with white hairs.

The maxillae and lip are dark brown. The maxillae are rather squarely truncated, their inner edges slanting to the lip, which is half as long as the maxillae, and pointed.

The sternum is dark brown, covered, as are the surrounding thighs, with white hairs. The anterior thighs are separated by more than the width of the lip at the base.

The relative length of the legs is 4, 3, 1, 2. The first and second pairs are a little stouter than the third and fourth. The first leg has the femur much enlarged on the under side, narrowing at the extremity, and the tibia with its extremity much wider than the proximal end of the metatarsus. In color the legs are brownish red, with much long whitish hair. There are femoral, tibial, and metatarsal spines on the four pairs.

The abdomen is light reddish brown, and is encircled by a white band. Near the apex two short transverse, parallel white bars, on each side, extend from the encircling band toward the middle of the abdomen; at about the middle point of the dorsum is a large white spot ; and near the base are two small indistinct white spots formed by a few short hairs.

The venter is covered with whitish hairs, but has a central, longitudinal, darker band.

Habitat, Iowa.

II. ATTUS AESTIVALIS, new.

(Plate I, Figures 2 to 2c.)

MALE Length, 1.8 mm.; width of abdomen, 1.1 mm.; length of ceph.th., 2.1 mm.; width of ceph.th., 1.5 mm.

Legs.	Fem.	Pat.	Tib.	Metat.	Tar.	Total.
1.	1.2	.8	1.1	.6	.4	4.1
2.	1.	.6	.7	.4	.3	3.
3.	1.	.5	.4	.4	.3	2.6
4.	1.3	.8	.9	.5	.4	3.8

The cephalothorax is low ; the thoracic part is longer than the cephalic ; the color is bright rufous, covered with short white hairs: when these hairs are rubbed off there are seen two black spots in the middle of the eye-region, and a black line on each side extending from the dorsal to the lateral eye.

The quadrangle of the eyes is wider than long, and not wider behind. The dorsal is as large as the lateral eye, and is situated higher on the cephalothorax, so that its lower border is in a line with the upper border of the lateral eye. The small median eye is equidistant from the dorsal and lateral eyes, and is in a line with the lower border of the dorsal eye. The anterior row of eyes is slightly curved, a line from the top of the middle eyes cutting the upper borders of the lateral eyes. The middle eyes are subtouching ; the lateral eyes are half as large as the middle eyes, and are slightly separated from them. The clypeus is two-thirds as wide as the middle eyes, retreating, covered with white hairs.

The palpus is pale ; the femur is long, the patella is longer than the tibia, and equally stout ; the bulb is large and complicated.

The falces are moderately long and stout, vertical, notched on the lower edge, and brown in color.

The maxillæ are squarely truncated, and are curved, the upper outer corners projecting. They are pale brown in color. The lip is less than half as long, and is darker.

The sternum is pointed behind, rounded in front, and brown in color. The anterior thighs are separated by more than the width of the lip at the base.

The relative length of the legs is 1, 4, 2, 3. The first pair is the stoutest. In color they are light rufous, with darker bars. There are femoral, tibial, and metatarsal spines on the four pairs ; those on the femoral joints are very small ; those on the metatarsus of the fourth leg extend to the base.

The abdomen is light pink, with a white basal band which extends on to the sides ; on the dorsum are two longitudinal rows of rufous dots, with four dots in each row. Between the rows is a succession of white ∧s, pointing forward.

The venter is pale pink, with a central longitudinal, reddish band ; it is covered with short white hairs.

FEMALE—Length, 6.2 mm. ; width of abdomen, 2.3 mm. Length of ceph.th., 2 mm. ; width of ceph.th., 1.5 mm.

Legs	Fem.	Pat.	Tib.	Metat.	Tar.	Total.
1.	1.2	.6	.7	.5	.5	3.5
2.	1.	.5	.5	.4	.3	2.8
3.	.8	.4	.6	.4	.3	2.5
4.	1.1	.8	1.	.6	.5	4.3

The palpus is pale, with white hairs.

The falces are shorter than in the male, and not notched.

The maxillæ are straighter than in the male, and not so squarely truncated. The lip is a little more than half as long.

The legs are sometimes like those of the male, and sometimes pale, barred with very dark rufous.

The abdomen is usually light yellowish, with two rows of rufous dots, as in the male, but it is sometimes dark brownish rufous, while on the dorsum are two rows of pale dots with five dots in each row ; the dots of the pair second from the apex are elongated into short transverse lines, and the dots of the posterior pair are sometimes united, forming one spot. On the sides are five oblique pale bars.

The venter is reddish, or brownish, with a central darker band, and is covered with short white hairs.

The epigynum presents a plate wider anteriorly, with a rounded opening at the posterior edge.

Habitat, Pennsylvania.

Observations. We are indebted for this species to Mr. Philip Nell.

III. ATTUS SPLENDENS, new.

(Plate I, Figures 3 to 3b.)

MALE Length, 6.5 mm.; width of abdomen, 2.2 mm.; length of ceph.th., 3 mm.; width of ceph.th., 2.2 mm.

Legs.	Fem.	Pat.	Tib.	Metat.	Tar.	Total.
1.	.9	.9	1.	.6	.5	3.9
2.	1.1	.7	.8	.6	.5	3.7
3.	1.3	.9	1.	1.	.7	4.9
4.	1.2	.8	1.	.9	.6	4.5

The cephalothorax is rather high ; the cephalic part is shorter than the thoracic, and slants a little toward the anterior eyes ; the anterior part of the thoracic region is prominent ; the posterior part slopes abruptly to the base. The fovea is deep. The whole cephalothorax is covered with dark, but highly iridescent scales, which have red, blue and green reflections. If the specimen is looked at in alcohol, two whitish testaceous spots appear just behind the dorsal eyes. The lower narrow margin is black. On the anterior part of the eye-region are some upright black hairs.

The quadrangle of the eyes is wider than long, and wider behind.

The dorsal eye is about as large as the lateral, and is placed higher, so that a line from its lower border cuts the upper border of the lateral eye. The small median eye is in a line with the middle of the dorsal eye, and is equidistant from the dorsal and lateral eyes. The anterior row of eyes is straight. The middle eyes are slightly separated ; the lateral eyes are separated from the middle eyes by about one-third their diameter. The middle are more than twice as large as the lateral eyes. The clypeus is nearly as wide as the middle eyes. It is inclined backward, and is darkly iridescent.

The palpus has the tarsus long and oval ; the tibia is shorter and narrower than the patella, is wider than long, and has a very strong, short apophysis, two-thirds as wide as long, which follows, and lies close to the side of the tarsus ; the patella is as wide as long, and a little wider at the distal end ; the femur is as long as the patella and tibia, including the apophysis, and has two strong spines.

The falces are not robust, vertical, scarcely as long as the face ; they are dark reddish brown, with long white hairs on their inner edges.

The maxillæ are twice as long as the lip ; they are straight on the inner side above the lip, and excavated below for the sides of the lip ; they are truncated at the extremity ; in color they are white at the extremity and on the inner edge, and otherwise reddish. The lip is reddish tipped with white ; it is wider than long, and a little narrower toward the tip, which is slightly hollowed.

The sternum is testaceous, darker than the thighs, with white hairs ; all the thighs are clothed with white hairs ; those of the anterior pair are separated by more than the width of the lip at the base. The sternum projects slightly beyond the anterior thighs.

The relative length of the legs is 3, 4, 1, 2. The first and second pairs are a little more robust than the third and fourth. They are dark, almost black in color, somewhat testaceous underneath. There are exinguinal, femoral, patellary, tibial and metatarsal spines on the four pairs.

The abdomen is bright iridescent red, lighter around the margin, and having a purplish tinge on the dorsum. There are four indented dots near the base, and sometimes a whitish basal band.

The whole abdomen is covered with sparse black hairs, and the spinnerets are black. When the spider is in alcohol the base and a curved oblique band on each side appear red, and the dorsum dark iridescent green, with a short longitudinal red band, narrowest in the middle.

The venter is iridescent red, with two indistinct, darker, longitudinal bands.

FEMALE—Length, 7.7 mm.; width of abdomen, 2.9 mm.; length of ceph.th., 3.1 mm.; width of ceph.th., 2.5 mm.

Legs.	Fem.	Pat.	Tib.	Metat.	Tar.	Total.
1.	2.1	1.	1.	.9	.7	5.7
2.	1.9	.9	.9	.4	.4	4.5
3.	2.6	1.5	1.5	1.	.6	7.2
4.	2.1	.8	1.	1.1	.8	5.8

The cephalothorax is not iridescent as in the male. The thoracic part has a black region at the base, from the anterior edge of which there comes off on each side a black band which curves forward and passes along the side of the cephalothorax below the dorsal and small median eyes, and reaches the anterior lateral eye. The black region at the base is usually limited in front, just posterior to the dorsal eyes, by a scalloped white or rufous band which curves forward in the middle. Anterior to this is sometimes another band, black, just between the dorsal eyes. The eye-region is rufous, or, more rarely, black. The sides of the cephalothorax below the black bands are covered with white hairs down to the lower narrow margin, which is black. The colors of the cephalothorax, particularly above, vary greatly between white, rufous and black.

The dorsal eye is placed so much higher than the lateral eye that a straight line from its lower border passes above the lateral eye. The small median eye is in a line with the lower edge of the dorsal eye, and is nearer to it than to the lateral eye.

The palpus is light rufous, or pale, covered with white hairs.

The maxillæ and lip are brown in color, short, and rounded.

The legs are varied with rufous and black. There are femoral, patellary, tibial, and metatarsal spines on the four pairs.

The ground color of the abdomen is velvety black; there is a white band at the base; a central longitudinal white band which begins at a little distance from the basal band and stops short of the apex, and which is usually slightly enlarged, or pointed like an arrow at its anterior extremity; and a scalloped white band on each side which begins in a line with, or a little posterior to, the beginning of the central band, and, curving downward, passes along the side to the apex. The white of these abdominal markings has usually a distinct tint of salmon-color.

The venter is covered with light brownish hairs.

The epigynum has the usual opening at the posterior edge. The tubes may be seen through the skin.

Habitat, Wisconsin.

IV. ATTUS OCTO-PUNCTATUS, new.

MALE—Length, 8 mm.; width of abdomen, 2.8 mm.; length of ceph.th., 3.1 mm.; width of ceph.th., 3 mm.

Legs.	Fem.	Pat.	Tib.	Metat.	Tars.	Total.
1.	2.6	1.1	2.1	1.7	.9	8.4
2.	1.9	.9	1.	1.	.8	5.6
3.	2.	.9	1.1	1.	.9	5.9
4.	2.2	1.	1.9	1.9	.9	7.9

The cephalothorax is high, the thoracic part being much longer than the cephalic, and dilated; the junction of the sides with the upper surface has no distinct edge, but is rounded. The upper surface is black, covered with short white hairs; the sides are dark brown with some irregular whitish spots, and sparse yellowish hairs; the lower margin is slightly darker than the rest of the sides; there are some long reddish yellow hairs on the eye-region.

The quadrangle of the eyes is wider than long, and wider behind. The dorsal eye is a little smaller than the lateral, and is placed much higher, a straight line from its lower border passing above the lateral eye. The very small median eye is in a line with the lower border of the dorsal eye, but is much nearer the lateral eye. The anterior row of eyes is slightly curved, a straight line from the top of the middle eyes cutting the upper borders of the lateral eyes. The middle are almost twice as large as the lateral eyes, and all four are well separated. The clypeus is two-thirds as wide as the middle eyes, vertical, dark brown, mottled with white, and having a few yellowish hairs.

The tibia of the palpus has a short thornlike apophysis; the bulb is large and thick, projecting backward to the extremity of the patella. The falces are rather stout, a little inclined forward, and dark brown in color. The maxillæ are short, abruptly truncated, and inclined toward the lip; the lip is short and pointed.

The sternum is dark brown, long, and oval.

The relative length of the legs is 1, 4, 3, 2. The first pair is the stoutest, the second next. They are brown, with brown hairs. There are spines on the tibial and metatarsal joints of the four pairs, those on the metatarsi of the fourth extending to the base.

The abdomen is black, covered with very short white hairs, and longer yellowish hairs. Near the base is a short curved white line; posterior to this are two longitudinal rows of white spots, four spots in each row; these are divided into two anterior and two posterior pairs, the four posterior spots being elongated obliquely. On the posterior sides are two or three short, oblique, white lines.

The venter is black, with short yellowish hair.

Habitat; Missouri.

Observations. This species was obtained through the kindness of the Rev. Henry C. McCook.

V. Attus hoyi, new.

(Plate 1, Figure 5.)

MALE. Length, 5.2 mm.; width of abdomen, 1.8 mm.; length of ceph.th., 2.3 mm.; width of ceph.th., 1.9 mm.

Legs.	Fem.	Pat.	Tib.	Metat.	Tar.	Total.
1.	1.7	1.	1.	.9	.6	5.2
2.	1.4	.8	.8	.6	.4	4.
3.	1.9	.8	.9	.8	.8	5.2
4.	1.4	1.	.9	.9	.7	4.9

The cephalothorax is high; the thoracic part is longer than the cephalic. The anterior part of the thoracic region, and the region of the eyes are rufous, a little mixed with black. Just posterior to the dorsal eyes is a transverse black band, wide at the sides, narrowing and curving forward a little in the middle. A white band on each side, beginning at or near the base, passes obliquely upward and forward for a short distance, and then widens and curves downward and forward along the side of the cephalothorax as far as the anterior lateral eyes, passing below the small median and dorsal eyes; another white band which almost joins this, passes over the top of the head just above the anterior eyes. Each of the anterior eyes is surrounded by hairs, which are white, excepting just between the eyes, where they are red. Looking from above, this gives the appearance of three minute red tufts. The base of the cephalothorax and wide regions on the sides below the lateral white bands are velvety black.

The quadrangle of the eyes is wider than long, and not wider behind. The dorsal eye is a little smaller than the lateral, and is placed a little higher, a straight line from its lower border cutting the lateral eye above the middle. The small median eye is near the dorsal eye, and is on a line with its upper edge. The anterior eyes are in a straight row, and are all touching. The middle are a little more than twice as large as the lateral eyes. The clypeus is more than half as wide as the middle eyes, vertical, black, with a few white hairs.

The patella of the palpus is a little longer than the tibia, and equally stout. The tibia has a long curved apophysis extending about half the length of the tarsus. On the inner border of the tibia are several long hair-like spines. The tarsus is oval, truncated at the tip, clothed with short white hairs. The bulb occupies the posterior three-fourths of the tarsus. The hook is long, and is curved on the inner anterior margin. There are spines on the femur, patella and tibia.

The falces are not robust, vertical, dark testaceous, with sparse, short light hairs.

The maxillæ and lip are brown, short and rounded.

The sternum is long, oval, black or testaceous, with black and white hairs, particularly around the edge. The anterior thighs are separated by the width of the lip at the base.

The relative length of the legs is 3, 1, 4, 2. The first and second pairs are slightly stouter than the third and fourth. All the legs are pale to the middle of the femur; beyond this point they have alternate bands of black, and pale brown or rufous. There are spines on the femoral, patellary, tibial, and metatarsal joints of the four pairs, those on the metatarsi of the fourth extending to the base.

The abdomen is black, with an encircling white band, and a central, longitudinal, rufous band, which varies in width and length, in which are some white marks. These marks vary, consisting sometimes of two short oblique lines near the base, which almost meet to form a ∧, posterior to these a ∧ with a much wider angle, and still nearer the apex two or three smaller ∧s, all pointing forward. In other specimens there are four short, oblique, white lines, two near the base, and two more widely separated, posterior to the middle of the dorsum. In some cases the rufous band is shortened into a spot, and the white marks are small and indistinct. The posterior part of the black portion of the abdomen sometimes forms three scallops, the middle one being the deepest, and extending to the very apex, thus dividing the encircling white band.

The venter is black, covered with white, or white and rufous hairs.

Habitat, Pennsylvania, Wisconsin.

VI. Attus flavus, new.

(Plate I, Figure 6.)

MALE—Length, 6.1 mm.; width of abdomen, 2 mm.; length of ceph.th., 2.6 mm.; width of ceph.th., 2.2 mm.

Legs.	Fem.	Pat.	Tib.	Metat.	Tar.	Total.
1.	2.	1.2	1.9	1.4	1.	7.5
2.	1.5	.9	1	1.	.8	5.2
3.	1.2	.9	.9	.9	.8	4.7
4.	1.6	.9	1.1	1.	.9	5.5

The cephalothorax is high; the thoracic part is longer than the cephalic, and is a little dilated. In color the cephalothorax is black, covered with short brownish-yellow hairs; the lower margin is black; when the spider is in alcohol a light spot appears behind the dorsal eyes. The fovea is wide and shallow.

The quadrangle of the eyes is wider than long, and slightly wider behind. The dorsal eye is a little smaller than the lateral, and is placed so much higher that a straight line from its lower border passes above the lateral eye. The small median eye is on a line with the lower border of the dorsal eye, but is nearer the lateral eye. The anterior row of eyes is slightly curved, a straight line from the top of the middle eyes cutting above the centre of the lateral eyes; the lateral eyes are placed further back on the face than the middle eyes; the lateral eyes are half as large as the middle eyes, and are slightly separated from them, the middle eyes being also slightly separated. The clypeus is scarcely one-fifth as wide as the middle eyes, vertical, with a few brownish white hairs.

The palpus is reddish, with white hairs; the patella is much longer than the tibia; the tibia has a curved, thorn-like apophysis. The tarsus and bulb are long.

The falces are iridescent green, vertical, long and rather stout.

The maxillæ are squarely truncated. The lip is pointed.

The sternum is oval, black, with white hairs.

The abdomen is covered with short, light yellow hair. There is a white band at the base; the centre of the dorsum has a wide, longitudinal black band, which has some brownish hairs down its middle line; the black on either side of these hairs is interrupted in three places by spots of yellow hair, the two anterior spots being at about the middle point of the dorsum, and the four posterior much nearer the apex. There are two white, oblique bands on each side.

Habitat, Pennsylvania.

Observations. This species resembles A. sinister of Hentz, but differs from it in the relative length of the legs, and in having much yellow, and no rufous, on the abdomen.

VII. Attus rusticolus, new.

(Plate 1, Figure 7.)

FEMALE—Length, 4.7 mm.; width of abdomen, 2.1 mm.; length of ceph.th., 2.4 mm.; width of ceph.th., 1.9 mm.

Legs.	Fem.	Pat.	Tib.	Metat.	Tar.	Total.
1.	1.2	.8	.9	.7	.6	4.2
2.	1.	.7	.8	.5	.4	3.4
3.	1.	.6	.7	.4	.3	3
4.	1.8	.7	1.2	.7	.5	4.9

The cephalothorax is rather high; the thoracic part is longer than the cephalic; the color is uniform tawny, caused by a close mixture of reddish, black, and gray hairs. The lower margin has a narrow line of black. In some specimens there are one or two indistinct, darker, transverse bands on the thoracic part, which begin at the sides and curve forward toward the eyes.

The quadrangle of the eyes is wider than long, and not wider behind. The dorsal is a little smaller than the lateral eye, and is placed higher, a straight line from its lower border passing above the lateral eye. The small median eye is in a line with the lower border of the dorsal eye, and is nearer to it than to the lateral eye. The anterior row of eyes is straight; the lateral eyes are half as large as the middle eyes, and are slightly separated from them; the middle eyes touch each other. The clypeus is two-thirds as wide as the middle eyes, retreating, and covered with red and gray hairs.

The palpus is reddish, covered with white hairs.

The falces are weak, vertical, dark brown in color, with a few reddish hairs at the base.

The maxillæ and lip are brown, paler at the tips. The maxillæ are rounded; the lip is about half as long, and is truncated.

The sternum is almost black, with some sparse, whitish hairs. The anterior thighs are separated by more than the width of the lip at the base.

The relative length of the legs is 4, 1, 2, 3. The first pair is a little the stoutest. In color they are testaceous, covered with short white hairs. There are black bands at the extremity of the femur, at the base and extremity of the tibia, and at the extremities of the metatarsus, and tarsus. There are spines on the tibial and metatarsal joints of the four pairs, those on the metatarsi of the fourth extending to the base.

The abdomen is of the same tawny color as the cephalothorax. The markings are regular, but indistinct; there are two lightish spots just in front of the middle point of the dorsum, the posterior edges of which are outlined by black bands which pass downward over the sides. Posterior to these are two larger spots of the same light color whose posterior edges are outlined by a single black band which curves forward between them, forming a ∨. This black band passes nearly around the circumference of the spots, but does not quite en-

circle them. From the lateral edges of this band, on each side, an
indistinct curved black band passes downward; and posterior to this,
and parallel with it, is another black band. Extending from the apex
to the posterior light spots, is a series of small black As, which point
forward.

The venter is tawny, with a wide central gray band, not reaching
the apex. Just at the apex is a short, traverse white band.

The epigynum presents a plate which is rounded anteriorly, with
the opening, as usual, at the posterior edge.

Habitat, Wisconsin.

VIII. Attus tibialis, new.

(Plate 1, Figures 8 to 8a.)

MALE—Length, 4.9 mm ; width of abdomen, 2 mm.; length of ceph.th., 2.1 mm.;
width of ceph.th., 1.7 mm.

Legs.	Fem.	Pat.	Tib.	Metat.	Tar.	Total.
1.	1.3	.8	1.2	.8	.7	4.8
2.	1.3	.7	.7	.5	.4	3.6
3.	1.	.6	.5	.1	.3	3.
4.	1.4	.8	1.	.8	.6	4.6

The cephalothorax is low; the thoracic part is longer than the
cephalic. In color it is black, with its upper surface covered with
brownish white hairs. The lower narrow margin is white.

The quadrangle of the eyes is wider than long, and a little wider
behind. The dorsal is about as large as the lateral eye, and is placed
a little higher, so that its lower border is on a line with the middle of
the lateral eye. The small median eye is on a line with the upper
margin of the dorsal eye, and is equidistant from it and from the
lateral eye. The anterior row is curved, a straight line from the top
of the middle cutting the lateral eyes through the middle. The middle eyes are touching, and are twice as large as the lateral eyes, which
are more widely separated. The clypeus is less than one-fifth as wide
as the middle eyes, and is deeply notched. It is covered with whitish
hairs.

The palpus has the tibia as wide as long, with a short curved apophysis ; the tarsus is long. The bulb is simple.

The falces are rather robust, and vertical. The fang is short.

The maxillæ are short and rounded, brown in color. The lip is
darker, half as long as the maxillæ, also rounded.

The sternum is twice as long as wide, dark brown. The anterior
thighs are separated by the width of the lip at the base.

The relative length of the legs is 4, 1, 2, 3. The first pair is the
stoutest, the second next. The femur of the first is reddish above,
black beneath. The patella is reddish, and much more slender than
the femur and tibia. The tibia is black, and is very stout. On the
under side of these three joints is a thick fringe of brownish white

hair, thickest on the tibia. The metatarsus and tarsus are slender, and rufous, except a black ring on the distal end of the tarsus. The posterior three pairs are reddish brown, varied with black, and have some short white hairs. There are femoral, tibial and metatarsal spines on the four pairs, those on the metatarsi of the third and fourth placed only in a circle at their extremities.

The abdomen is black, nearly covered with small scales of a dull gold color. At the base are some short white hairs.

The venter is black.

FEMALE—Length, 5.7 mm.; width of abdomen, 2.3 mm.; length of ceph.th., 2.6 mm.; width of ceph.th., 2 mm.

Legs.	Fem.	Pat.	Tib.	Metat.	Tar.	Total.
1.	.9	.5	.8	.5	.4	3.1
2.	.9	.4	.7	.4	.4	2.8
3.	1.	.5	.8	.5	.5	3.5
4.	1.2	.7	1.	.9	.5	4.3

The clypeus is wider than in the male, being about one-fourth as wide as the middle eyes.

The abdomen is black, sparsely covered with short white hairs, and encircled by a white band. Near the apex, on each side, four or five short oblique white bands pass upward and forward from the encircling band, but do not meet in the middle. Low on the sides are golden scales, as in the male.

The epigynum presents a plate as wide as long, with rounded corners. Near the anterior border are two transverse dark spots. The opening is at the posterior edge.

Habitat, Wisconsin.

IX. ATTUS AGRESTIS, new.

(Plate 1, Figures 9 to 9a.)

FEMALE—Length, 10 mm.; width of abdomen, 3.6 mm.; length of ceph.th., 3.9 mm.; width of ceph.th., 3.1 mm.

Legs.	Fem.	Pat.	Tib.	Metat.	Tar.	Total.
1.	2.	1.4	1.5	.9	.8	6.6
2.	2.	1.2	1.3	.8	.7	6.
3.	2.1	1.2	1.7	1.4	.9	7.
4.	2.3	1.	1.9	1.3	1.	7.5

The cephalothorax is high, sloping abruptly behind; the cephalic and thoracic parts are of equal length. The color is pale yellowish, covered with short light hairs; the cephalic and thoracic parts are separated by a V-shaped band of reddish hairs, which is edged by a black line. The eyes are situated on black spots, this background bringing them out and making them very conspicuous. There are a few long, stiff, black hairs on each side of the eye area.

The quadrangle of the eyes is a quarter wider than long, and wider behind, the cephalothorax dilating beyond the dorsal eyes. The dorsal eye is as large as the lateral, and is in a line with it; the small median eye being nearer the lateral eye, on a line with its upper border, and situated on a low elevation. The anterior row of eyes is slightly curved, a straight line from the top of the middle eyes cutting the lateral eyes above the center. The middle eyes are subtouching; the lateral are more widely separated, and are less than half as large. The clypeus is about one-half as wide as the middle eyes, and is slightly inclined; it is covered with long yellow hairs, which also surround the middle eyes.

The palpus is slender and pale.

The falces are long, vertical, rather slender (being but little wider than the middle eyes), and pale. The fangs are weak.

The maxillae are enlarged and rounded at the extremities.

The sternum is oval, long and pale. The anterior thighs are separated by more than the width of the lip at the base.

The relative length of the legs is 4, 3, 1, 2; they are short in comparison with the size of the spider. The first and second pairs are stouter than the third and fourth. They are pale, with slightly darker rings on the distal ends of the joints. There are very weak femoral, tibial and metatarsal spines on the four pairs.

The abdomen is pale yellow, with small black dots and hairs irregularly disposed over the back and sides, leaving an unspotted median longitudinal narrow band. The dots and hairs are more numerous toward the base. The spinnerets are pale and slender.

The venter is pale and unspotted.

The epigynum presents a large dark spot and two overlapping plates on each side. The opening, at the posterior edge, is long and inconspicuous.

Habitat, Pennsylvania.

Observations. For this species we are indebted to Mr. Philip Nell.

X. ATTUS ARIZONENSIS, new.

(Plate II, Figures 10 to 10a.)

MALE—Length, 11 mm.; width of abdomen, 3 mm.; length of ceph.th., 4.9 mm., width of ceph.th., 4.2 mm.

Legs.	Fem.	Pat.	Tib.	Metat.	Tar.	Total.
1	3	2.1	2.8	2.1	1.4	11.4
2	2.7	1.7	1.9	1.6	1.1	9.
3	2.8	1.9	1.7	1.5	.9	9.
4	3.2	1.7	2.1	2.	1.	10.3

The cephalothorax is high; the cephalic and thoracic parts are equal in length; the thoracic part slopes abruptly from the depression which divides it from the cephalic part. This depression is deep and convex anteriorly. The whole cephalothorax is enlarged both above and on the sides, in the middle. The sides slope inward toward the lower margin. The color is velvety black, with two wide white lateral bands beginning just before and below the dorsal eyes, and almost meeting in the middle of the thorax behind the depression. There is a band of grayish brown hair above the anterior eyes.

The quadrangle of the eyes is slightly longer than wide, and markedly wider behind. The dorsal is larger than the lateral eye, and is in a straight line with it. On the sides, just in front of the dorsal eyes, there is a marked projection or cheek, so that the dorsal eyes are almost hidden from in front. The small median eye has its lower edge on a line with the upper edge of the lateral eye, and is much nearer to it than to the dorsal eye. The anterior row is slightly curved, a straight line from the summit of the middle eyes cutting through the centre of the lateral eyes. The middle eyes are touching, and are twice as large as the lateral eyes, which are separated from the middle eyes by nearly their diameter. The face is so curved that the lateral are set further back than the middle eyes. The clypeus is as wide as the middle eyes, and vertical. White hairs cover it, and surround the middle eyes.

The palpus has the tibia and patella nearly square, the tarsus oval, and covered with short yellow hairs. The bulb is not complicated.

The falces are stout, long, and vertical. The fang is short. The anterior surface is thinly covered with thin white hairs.

The mouth parts are black. The maxillæ are narrow at the base; enlarged at the tip; the upper edge is rounded, but the corners are sharp. The tongue is about half as long, and rounded. The sternum is black, about as wide as the intermediate thighs, and deep set. The anterior thighs are much the thickest and longest, and are separated by the width of the lip at the base.

The relative length of the legs is 1, 4, 2, 3 ; they are about equal in thickness, but the femur of the first is slightly stouter than the others. In color the legs are yellowish excepting the femur of the first, which is black above and pale below ; the third and fourth pairs are darker than the first and second, and show some reddish rings on the patellæ and tibiæ. All the legs have, on the under side, long fine yellow hair, which is longest on the first pair. There are femoral, patellary, tibial, and metatarsal spines on the four pairs, those on the metatarsi of the fourth extending to the base.

The abdomen is long and narrow, wide at the base, and pointed at the apex. The dorsum is much below the plane of the cephalothorax. The color is light brown ; behind the middle is a median, longitudinal, velvety black band. At the apex the abdomen is truncated, and the spinnerets are turned downward in the same direction as the truncated face of the abdomen, and have the same velvety dark color. In the middle of the abdomen is a pair of impressed dots, and a second pair, just in front of these, is very indistinct. At the apex are two white spots, one on each side of the black line. On the upper sides of the abdomen there is a black line extending to the apex; the under sides have wide white bands, formed of white hairs directed downward, which extend beneath on to the venter.

The venter is velvety black, darkest behind, the white bands marking it off into a long triangle, the apex in front of the spinnerets. Habitat, Arizona.

Observations. For this interesting species we are indebted to the Rev. Henry C. McCook.

XI.　Attus miniatus, new.

(Plate II, Figures 11 to 11a.)

FEMALE—Length, 13 mm.; width of abdomen, 4 mm.; length of ceph.th., 5.9 mm.; width of ceph.th., 5 mm.

Legs.	Fem.	Pat.	Tib.	Metat.	Tar.	Total.
1.	3.	3.	2.5	1.9	1.1	11.5
2.	2.6	2.	2.	1.9	1.	9.5
3.	3.1	1.9	1.9	1.9	1.9	10.7
4.	3.9	2.2	2.9	2.1	1.9	13.

The cephalothorax is high; the thoracic part is longer and wider than the cephalic. The eye region is covered with short, bright red hairs, intermixed with long black hairs, the latter forming small tufts between the small median and dorsal eyes. The thoracic part and sides are covered with coarse yellowish white hairs. The lower narrow margin is black.

The quadrangle of the eyes is wider than long, and wider behind. The dorsal eye is as large as the lateral, and on a line with it; the small median eye is on a line with their upper borders, and nearer the lateral than the dorsal eye. The anterior row of eyes is slightly curved, a line from the summit of the middle eyes cutting the lateral eyes above the centre. The middle eyes are twice as large as the lateral. All the anterior eyes are separated, the lateral more widely than the middle eyes. The clypeus is two-thirds as wide as the middle eyes, and is covered with dense whitish hair.

The palpus is pale, tipped with black, and is covered with long white hairs.

The falces are short, stout and vertical; they are dark colored, but highly irridescent, with red, purple, yellow and green reflections. The fang is long.

The maxillæ are rounded, brown, with the inner edge, which slopes to the lip, pale.

The sternum is dark brown; pointed in front, rounded behind.

The relative length of the legs is 4, 1, 3, 2. The first pair is the stoutest, the second next. They are black, excepting the metatarsi, which are pale, and are entirely covered with long whitish hair.

There are stout femoral, tibial and metatarsal spines on the four pairs, those on the metatarsis of the fourth extending to the base.

The abdomen is covered at the base with coarse gray hairs; otherwise it is covered with short, bright red and long whitish hairs. On the middle of the dorsum are three eye spots, a large one in the center, and posterior to this, two, which are transversely elongated. They are of the same bright red color, encircled by black rings. The sides are covered with long gray hairs. The spinnerets are black.

The venter has a wide central longitudinal band, black, limited by the gray hair which comes low on the sides. Just at the apex, between the black band and the black spinnerets, is a narrow transverse band of gray hairs.

The epigynum presents a rounded plate, wider than long, with the opening at the posterior edge, which is notched.

Habitat, Florida.

XII. ATTUS M'COOKII, new.

(Plate 11, Figure 12.)

FEMALE—Length, 15.4 mm.; width of abdomen, 6 mm.; length of ceph.th., 4.9 mm.; width of ceph.th., 4.2 mm.

Legs.	Fem.	Pat.	Tib.	Metat.	Tar.	Total.
1.	3.7	1.9	2.1	1.6	1.3	10.7
2.	3.	1.2	1.5	1.3	1.	8.
3.	3.1	1.2	1.5	1.3	·1.3	8.2
4.	3.7	1.8	2.5	1.8	1.2	11.

The cephalothorax is high; the thoracic part is longer than the cephalic. In color it is dark rufous covered with tawny hair.

The quadrangle of the eye is wider than long, and much wider behind. The dorsal eye is as large as the lateral eye, and is on a line with it. The small median eye is a little higher than the upper border of the lateral eye, and is much nearer the lateral than the dorsal eye. The anterior row of eyes is slightly curved, a straight line from the top of the middle eyes cutting the upper borders of the lateral eyes. The middle eyes are twice as large as the lateral, and are sub-touching; the lateral are separated from the middle by about half their diameter. The clypeus is about one-half as wide as the middle eyes; it is vertical, and is dark rufous in color, with long white hairs.

The palpus is rufous, with white hairs.

The falces are not very stout, slightly inclined backward, a little longer than the face; their color is dark rufous.

The maxillæ have their outer edges slanting outward; the extremities are rounded; the inner edges slant toward the lip; the lip is rounded at the tip, half as long as the maxillæ. In color the mouth parts are rufous, with the tip of the lip and the inner edges of the maxillæ pale.

The sternum is rufous; the anterior thighs are separated by only the width of the lip at the base.

The relative length of the legs is 4, 1, 2, 3; the first is much the stoutest. They are dark rufous in color, with blackish bars.

The abdomen is golden yellow, with short hairs of the same color. There are four indented dots near the base, and posterior to these two indistinct dark bands extend to the apex. The sides are creased. The abdomen is distended with eggs in the only specimen which we have of this species.

The venter is yellow, with three indistinct dark bands.

The epigynum presents a rounded plate with a large opening at the posterior edge. The internal parts are visible through the integument.

Habitat. The only specimen which we have of this species is from the United States, but the exact locality is unknown.

Observations. We have named this species for the well-known naturalist, Rev. Henry C. McCook, of Philadelphia.

XIII. ATTUS PEREGRINUS, new.

(Plate II. Figures 13 to 13a.)

MALE—Length, 5.3 mm.; width of abdomen, 1.9 mm.; length of ceph.th., 2.8 mm; width of ceph.th., 2 mm.

Legs.	Fem.	Pat.	Tib.	Metat.	Tar.	Total
1.	1.5	1.	1.6	.7	.5	5.3
2.	1.4	.9	.9	.8	.5	4.5
3.	2.1	1.2	1.2	1.	.9	6.4
4.	2.	.7	1.	1.2	.7	5.6

The cephalothorax is high, especially so near the middle of the thoracic part; in front it inclines toward the face, and behind falls abruptly to the base. The thoracic is about one-fourth longer than the cephalic part. In color the cephalothorax is black, with the eye region covered with short dark brown hairs, bordered in front and behind by a light yellowish edge. There are a very few short white hairs on the thorax. The fovea is rather deep, and is convex in front.

The quadrangle of the eyes is wider than long, and not wider behind. The dorsal eye is a little larger than the lateral eye, and in a line with it. The small median eye is slightly nearer the dorsal than the lateral eye, and is on a line with their upper borders. The anterior eyes are surrounded by white rings, those around the middle eyes being formed by long hairs. They are in an almost straight row, a straight line from the top of the middle eyes, cutting the upper borders of the lateral eyes. The middle are about twice the diameter of the lateral eyes, and are sub-touching. The lateral are separated from the middle eyes by nearly their own diameter. The lateral are placed further back on the face than the middle eyes. The clypeus is retreating, and

3

is about three-fourths the diameter of the middle eyes. It is densely covered with rather long white hairs.

The palpus is slender, except the tarsus, which is large and oval; the dorsal surface is covered with short white hairs. The patella is longer than the tibia. There is a short tibial apophysis.

The falces are rather weak; they are about as wide as the middle eyes, and are nearly vertical; the anterior surface is covered with long white hairs. The fangs are small.

The maxillae and lip are pale, but a little darker than the sternum and legs. The maxillae are rounded; the lip about half their length.

The sternum is pale, with white hairs. It is pointed behind, wide in the middle, and truncated in front. The anterior thighs are the largest, and are separated by the width of the lip at the base.

The relative length of the legs is 3, 4, 1, 2, the first pair being the stoutest. In color they are pale yellow, with some short white hairs and a few longer black ones. The first pair are darker toward the base. The patella of the third gradually widens towards the distal end, where it projects over the tibia. This widened portion bears a short pale spine, which projects over the tibia, just behind which, on the anterior face, is a small black dot. There are slender femoral, tibial and metatarsal spines on the four pairs, extending to the base on the metatarsi of the fourth.

The abdomen is small and is pointed behind. The base and sides are covered with white hairs ; the dorsum is black, perhaps originally also covered with white hairs. There are two pairs of indented dots near the middle.

The venter is covered with white hairs.

Habitat, Connecticut.

Observations. We are indebted for this specimen to Mr. H. Van Rensselear, of New York.

XIV. ATTUS PRINCEPS, new.

(Plate 11, Figure 11.)

FEMALE. Length, 8.1 mm.: width of abdomen, 3 mm.; length of ceph.th., 3.9 mm.; width of ceph.th., 3.1 mm.

Legs.	Fem.	Pat.	Tib.	Metat.	Tars.	Total.
1.	2.	1.5	1.7	1.	.5	7.1
2.	1.9	1.1	1.1	1.	.8	5.9
3.	2.	1.	1.	1.2	.8	6.
4.	2.5	1	1.6	1.4	1.	7.5

The cephalothorax is high; the thoracic part is longer than the cephalic; in color it is dark rufous, black in the region of the eyes covered with short yellowish white hair, and having some long black hairs around the eyes. The lower narrow margin is black.

The quadrangle of the eyes is wider than long, and much wider

behind. The dorsal is as large as the lateral eye, and is on a line with
it; the small median eye is just above their upper borders, and is
slightly nearer the lateral eye. The anterior row of eyes is slightly
curved, a straight line from the top of the middle eyes cutting the
upper borders of the lateral eyes. They are all well separated, the
lateral eyes the more widely. The middle are twice as large as the
lateral eyes. The clypeus is one-half as wide as the middle eyes. It
is vertical, and is clothed with dense white hairs.

The palpi are rufous, with white hairs.

The falces are short, stout, and vertical. In color they are iridescent green.

The maxillæ are thick, stout and rounded. In color they are
reddish, tipped with white.

The relative length of the legs is 4, 1, 3, 2. The first pair is the
stoutest, the second next. In color they are distinctly reddish, growing darker toward the extremities, with whitish hairs. There are
spines on the femoral, tibial and metatarsal joints of the four pairs,
those on the metatarsi of the fourth extending to the base.

The sternum is rounded; black, with white hairs.

The abdomen is rather pointed. It is black, covered with coarse
tawny hairs, with which a few gray hairs are intermixed. These gray
hairs predominate around the base, giving that region a whitish appearance, and are a little thicker on either side of the middle point of
the abdomen, making two grayish spots.

The venter is blackish, with coarse gray hairs.

The epigynum presents a rounded plate with a large square opening at the posterior edge. The curved tubes are visible through the
skin in the anterior half.

Habitat, Pennsylvania.

XV.   ATTUS QUADRILINEATUS, new.

(Plate II, Figure 15.)

FEMALE—Length, 5.5 mm.; width of abdomen, 2 mm.; length of ceph.th., 2 mm.; width
of ceph.th., 1.5 mm.

Legs.	Fem.	Pat.	Tib.	Metat.	Tar.	Total.
1.	1.2	.8	.8	.4	.2	3.4
2.	1.	.5	.5	.2	.2	2.4
3.	1.	.4	.5	.2	.2	2.2
4.	1.2	.5	1.	.8	.4	3.9

The cephalothorax is low and flat; the thoracic part is longer than
the cephalic. The eye region is black, covered with yellowish hair;
the thoracic region is brownish testaceous, with sparse yellowish hair.
The lower narrow margin is black.

The quadrangle of the eyes is wider than long, and wider behind.
The dorsal eye is as large as the lateral, and is prominent; it is so

placed that a straight line from its lower border cuts the upper border of the lateral eye. The small median eye is on a line with the upper border of the dorsal eye, and is slightly nearer to it than to the lateral eye. The anterior row of eyes is very slightly curved, a line from the summit of the middle cutting the lateral eyes through their upper borders. The clypeus is about one-quarter as wide as the middle eyes, vertical, dark colored, with a few long white hairs.

The palpus is pale, testaceous, with hair, light brownish, except at the extremity, where it is white.

The falces are weak, vertical, short, light brown in color.

The maxillæ are narrow, and slightly inclined outward, rounded at tip. The lip is half as long, and rounded.

The sternum is pointed behind and rounded in front. The anterior thighs are separated by less than the width of the lip at the base.

The relative length of the legs is 4, 1, 2, 3. The first pair is the stoutest and darkest, and has the femoral joints flattened. The other legs are light brownish, testaceous with black hairs. There are femoral, tibial and metatarsal spines on the four pairs. Those on the first pair are long and stout ; those on the third and fourth weak, and found only at the extremity of the metatarsi.

The abdomen is blackish, with sparse yellow hairs. Four narrow lines of white hairs begin at the base ; the median two start from the same point, and separating a little, pass, one on each side of the middle, to the apex, terminating on either side of the spinnerets. The lateral lines extend along the sides from base to apex. They are not visible from above.

The venter is blackish, with short, sparse white hairs.

The epigynum presents a wide heart-shaped plate, with the apex in front. The opening is large and on the posterior edge. The tubes may be distinctly seen through the skin.

Habitat, Pennsylvania, Wisconsin.

XVI.   ATTUS PINUS, new.

(Plate II. Figure 16.)

FEMALE Length 5.9 mm.; width of abdomen, 2.2 mm.; length of ceph.th., 2.5 mm.; width of ceph.th. 2 mm.

Legs.	Fem.	Pat.	Tib.	Met.	Tar.	Total.
1.	1.5	1.	1.2	.5	.4	4.6
2.	1.5	.9	.9	.5	.4	4.2
3.	1.8	.9	1.	.8	.7	5.2
4.	2.	1.	1.1	1.	.9	6.

The cephalothorax is low, the thoracic part being longer than the cephalic; the eye region is black, with reddish brown hairs; the thoracic part and sides are reddish black; behind the dorsal eyes is a transverse band of gray and reddish hairs, which passes forward and downward onto the sides, extending below the dorsal, median and lateral eyes.

The quadrangle of the eyes is wider than long, and not wider behind. The dorsal eye is a little smaller than the lateral eye, and is placed so much higher that a straight line from its lower border passes above the lateral eye. The small median eye is on a line with the dorsal eye. The anterior eyes are in a straight row, and are subtouching; the lateral are more than half as large as the middle eyes. The clypeus is one-quarter as wide as the middle eyes, retreating, covered with white hairs.

The palpus is long and dark colored, with black and white hairs.

The falces are not robust, slightly inclined backward, dark reddish, with a few white hairs.

The maxillæ and lip are short; the lip is square, the maxillæ rounded. In color they are reddish, tipped with white.

The sternum is large and oval; dark at the margin and paler in the centre. The anterior thighs are separated by more than the width of the lip at the base.

The relative length of the legs is 4, 3, 1, 2, the first pair being the stoutest. They vary from light brown to black, the tarsi being pale. There are many spines on the femoral, tibial and metatarsal joints of the four pairs, those on the metatarsi of the fourth extending to the base.

The abdomen is rufous, or light brown, with a grayish basal band, just back of which is a very much curved but not acutely pointed black band, which passes diagonally downward over the sides. From the middle point of this band, a slender, black, longitudinal line extends backward for a short distance; beginning on a line with its termination are two short diagonal black bands, one on each side; and posterior to, and partly included within these, are two more, which reach to either side of the apex. From the apex a short, slender, longitudinal black line extends a little way forward. The sides are grayish rufous.

* The venter is dark gray, with some whitish longitudinal lines.

The epigynum has an opening at the posterior edge. The internal parts are visible through the integument.

Habitat, Wisconsin.

XVII. Attus johnsonii, new.

(Plate 11, Figures 17 to 17a.)

MALE. Length, 10 mm.; width of abdomen, 3.2 mm.; length of ceph.th., 5.1 mm., width of ceph.th., 3.3 mm.

Legs.	Fem.	Pat.	Tib.	Metat.	Tar.	Total.
1.	2.9	2.1	2.7	1.8	1.1	10.6
2.	2.7	1.7	1.7	1.5	1.	8.6
3.	2.7	1.6	1.6	1.9	1.1	8.9
4.	3.6	1.8	2.2	2.	1.3	10.9

The cephalothorax is moderately high ; the thoracic part is longer than the cephalic. In color it is jet black, with some long black hairs in the eye region.

The quadrangle of the eyes is wider than long, and wider behind. The dorsal is as large as the lateral eye, and in a line with it. The small median eye is just above their upper borders, and is nearer the lateral eye. The anterior row of eyes is curved, a straight line from the top of the middle eyes cutting the lateral eyes above the centre. The lateral are half as large as the middle eyes, from which they are separated by nearly their diameter. The middle eyes are well separated. The clypeus is one-half as wide as the middle eyes, vertical, dark brown, with a short fringe of white hairs at the edge.

The palpus has the bulb greatly enlarged behind. The hook is small.

The falces are stout, vertical, dark colored, but somewhat iridescent.

The maxillae are stout, thick, truncated, brown, with the inner edge which slopes to the lip, pale. The lip is half as long, rounded, brown.

The sternum is black with white hairs. The anterior thighs are separated by the width of the lip at the base.

The relative length of the legs is 4, 1, 3, 2. The first pair is the stoutest. In color they are dark rufous, barred with black. The tibia and patella of the first leg are equal in length. There are femoral, tibial and metatarsal spines on the four pairs, those on the metatarsi of the fourth being found only at the extremity.

The abdomen is bright red, with a tinge of yellow. The hairs are easily rubbed off, showing the integument to be jet black.

The venter is dark brown with white hairs.

FEMALE. Length, 8.5 mm.; width of abdomen, 3.3 mm.; length of ceph.th., 3.5 mm.; width of ceph.th., 2.8 mm.

The palpus is reddish, with white hairs.

The relative length of the legs is 4, 1, 2, 3. The tibia of the first is longer than the patella.

The abdomen has a whitish band at base, and a central longitudinal black band from the middle of the dorsum to the apex, in which are two opposed pairs of white dots.

The epigynum presents a plate which is curved in front, and narrow behind, with a large opening at the posterior edge.

Habitat, Washington Territory.

Observations. This species is named for Prof. O. B. Johnson, of Washington University, Seattle, W. T., to whose kindness we are indebted for an interesting collection of Washington Territory arachnida.

XVIII.　ATTUS FORMOSUS, new.

(Plate II, Figure 18.)

FEMALE　Length, 8.4 mm.; width of abdomen, 2.9 mm.; length of ceph.th., 4 mm.; width of ceph.th , 2.9 mm.

Legs.	Fem.	Pat.	Tib.	Metat.	Tar.	Total.
1.	2.3	1.5	1.5	1.	.7	7.
2.	1.9	1.1	1.	.7	.5	5.2
3.	2.	.9	.8	.7	.5	4.9
4.	2.8	1.2	1.5	1.1	.9	7.5

The cephalothorax is wide, and moderately high; the thoracic part is longer than the cephalic. It is black, entirely covered with short whitish brown hairs, and longer, sparse, black hairs, the black hairs all pointing forward, thickest just above the anterior eyes, and forming two tufts on each side, one just back of the lateral eye, and one between the small median and the dorsal eye. The sides are reddish. The lower narrow margin is white.

The quadrangle of the eyes is a third wider than long, and much wider behind. The dorsal is as large as the lateral eye, and is on a line with it. The small median eye is on a line with their upper borders, and is nearer the lateral eye. The anterior row of eyes is slightly curved, a straight line from the top of the middle, cutting the upper borders of the lateral eyes. The lateral are placed further back on the face than the middle eyes. The middle eyes are slightly, the lateral more widely, separated. The lateral are a little more than half as large as the middle eyes. This row of eyes is placed below the edge of the cephalothorax. The clypeus is one-half as wide as the middle eyes and vertical, the upper part is reddish and almost bare; the lower edge is fringed with thick, short, snowy white hair, which falls over the base of the falces.

The palpus is long and pale, with white hairs.

The falces are very robust, vertical, and moderately long. Their color is bright iridescent green.

The maxillæ are short, thick and rounded, light brown in color. The lip is half as long, rounded, and also light brown.

The sternum is brownish, somewhat iridescent, with white hairs.

The relative length of the legs is 4, 1, 3, 2. The first pair is the stoutest. In color they are reddish, with some white and black hairs. There are tibial and metatarsal spines on the four pairs, those on the metatarsi of the fourth extending to the base.

The abdomen is bright yellowish scarlet. A wide central longitudinal black band begins at the middle point, and extends backward to the spinnerets. Its anterior margin is marked by a short, curved, white band. On the edges of the black band are two pairs of white dots; the posterior two dots are close to the apex, and mark the posterior border of the red color. At the base of the abdomen is a wide white band, which extends on to the sides; and below this a black band passes around the sides, blending, at the apex, with the central black band. As the anterior ends of the encircling band are not visible from above, the black marking at the posterior part of the abdomen has the appearance of an anchor.

The venter is black, with white hairs, which form four indistinct lines, two at the edges, and one on either side of the center. The middle lines do not reach the apex.

The epigynum is not uncovered.

Habitat, Iowa.

Observations. This species is from the Putnam collection.

XIX. ATTUS ALBO-IMMACULATUS, new.

(Plates II and III. Figures 19 to 19a.)

FEMALE—Length, 6.1 mm.; width of abdomen, 2 mm.; length of ceph.th., 2 mm. width of ceph.th., 1.6 mm.

Legs.	Fem.	Pat.	Tib.	Metat.	Tar.	Total.
1.	1.1	.5	.6	.5	.5	3.2
2.	.8	.4	.5	.6	.3	2.6
3.	.9	.4	.4	.4	.3	2.4
4.	1.2	.7	.7	.9	.4	3.9

The cephalothorax is low. The cephalic and thoracic parts are equal in length. The color of the cephalothorax is white, being densely covered with short white hairs.

The quadrangle of the eyes is as long as wide, and wider behind. The dorsal eye is slightly larger than the lateral, and is placed lower on the caput, so that a straight line from the upper border of the dorsal cuts the lower border of the lateral eye. The small median is slightly higher than the lateral eye, and is nearer the lateral than the dorsal eye. The anterior row of eyes is slightly curved, a line from the top of the middle cutting the lateral eyes above the center. The middle are more than twice as large as the lateral eyes and are subtouching. The lateral are separated from the middle eyes by about their own diameter, and are placed further back on the face. The clypeus is half as wide as the middle eyes, vertical, and covered with dense short white hairs.

The palpus is covered with white hairs.

The falces are relatively stout, vertical, and have the anterior surface slightly rounded. They are covered with white hairs.

The maxillæ are enlarged toward the ends, and rounded. The lip is half as long as the maxillæ, and pointed.

The sternum is truncated in front, and pointed at the posterior extremity, which is in front of the posterior thighs. It is covered with white hairs. The anterior thighs are separated by less than the width of the lip at the base.

The relative length of the legs is 4, 1, 2, 3. The first pair is the stoutest. The femur of the first has a row of short and strong black hairs near the middle of the lower border. The patella is enlarged at the distal end, and is covered with white hairs, separated into two circles by an indistinct hairless median part. The tibia is thick; on the posterior half of the upper border is a ridge of short strong black hairs, and on the whole of the lower border a similar ridge, the hairs of this ridge having a length about equal to the thickness of the articulation; the anterior face of the tibia is black. The metatarsus and tarsus are much more slender than the other articulations. The other legs are covered with white hairs, with paler, more or less hairless longitudinal lines, most marked on the posterior two pairs. There are very weak femoral, tibial and metatarsal spines on the first and second pairs. The third and fourth pairs are unarmed. The femoral joints of the third and fourth have some stout black hairs.

The abdomen is slightly flattened and is in the same plane as the cephalothorax. It is densely covered with short white hairs, with a few longer, fine black hairs scattered over it.

The venter is covered with short thick white hairs.

The epigynum presents a large plate, wide behind, narrower and rounded in the front, with the opening at the posterior edge.

Habitat, Iowa.

Observations. This species is from the Putnam collection.

XX. ATTUS PALUSTRIS, new.

(Plate III, Figures 20 to 20c.)

MALE—Length, 5 mm.; width of abdomen, 2 mm.; length of ceph.th., 2.7 mm.; width of ceph.th., 1.6 mm.

Legs.	Fem.	Pat.	Tib.	Metat.	Tar.	Total.
1.	1.3	.9	1.	.8	.6	4.6
2.	1.1	.7	.7	.6	.5	3.6
3.	1.	.4	.5	.5	.5	3.4
4.	1.4	.7	1.	.8	.5	4.4

The cephalothorax is high; the thoracic part is longer than the cephalic. The color is dark brown with a reddish tinge towards the eyes, and nearly black at the base. Three white lines begin at the base and run forward; the median line ends before reaching the anterior eyes; the lateral lines pass just outside the dorsal eyes and inside the small median eyes, and are connected by a transverse white line which crosses the eye region just above the anterior eyes. There is sometimes a less distinct transverse white line behind the dorsal eyes. The sides of the cephalothorax have a narrow white line at the margin formed by hairs which seem to be combed upward.

The quadrangle of the eyes is wider than long, and slightly wider behind. The dorsal is as large as the lateral eye, and is placed higher on the caput, so that a straight line from its lower border passes above the lateral eye. The small median eye is equidistant from them, and is on a line with the middle of the dorsal eye. The anterior row of eyes is straight. The middle eyes are touching, and are more than twice as large as the lateral, which are separated from the middle eyes by about half their diameter. The clypeus is one-half as wide as the middle eyes, and is inclined. There is a white line just below the anterior eyes ; otherwise the clypeus is brown, and without hairs, excepting at the base of the falces, where there is a membranous ridge covered with white hairs.

The palpus has white hairs excepting on the tarsus. The tibia is rounded on the outer edge, but has no apophysis. The bulb is simple.

The falces are not robust ; they are long and vertical, of the same color as the clypeus, without hairs.

The maxillæ and lip are rounded.

The sternum is wide and oval ; in color, brown with white hairs. The anterior thighs are separated by more than the width of the lip at the base.

The relative length of the legs is 1, 4, 2, 3, the first pair being slightly the stoutest. They are brown with whitish bars. There are femoral, tibial and metatarsal spines on the four pairs, those on the metatarsus of the fourth leg extending to the base.

The abdomen is small, and pointed behind. The general color is brown ; there is a central longitudinal lighter band (sometimes wanting), on either side of which the abdomen is black. In the center of the abdomen are two transverse white bars, or large spots, one on each side. In front of these bars are one or two very small spots on each side. Behind the median bars are two indistinct light dots, one behind the other, in the middle line. Just above the apex are two white bars corresponding to the middle bars, but narrower. The sides are pale yellow.

The venter is light brown, with indistinct longitudinal dark bands.

FEMALE—Length, 5.3 mm.; width of abdomen, 2.4 mm.; length of ceph.th., 2.1 mm.; width of ceph.th., 1.8 mm.

Legs.	Fem.	Pat.	Tib.	Metat.	Tar.	Total.
1.	1.	.5	.6	.5	.4	3.
2.	.9	.5	.5	.4	.4	2.7
3.	.8	.4	.5	.4	.4	2.5
4.	1.3	.5	1	.7	.7	4.2

The cephalothorax is covered with mixed black and rufous hairs, the whole appearance being reddish. Between the dorsal eyes is a black transverse band. At the base is a black spot. Extending from near the base to a point in front of the dorsal eye is a light colored median, longitudinal line, which is crossed, behind the dorsal eyes, by an indistinct short transverse light line. On each side, passing below the dorsal eyes, is an indistinct, light, longitudinal line. The lower margin has a narrow line of black, just above which is a narrow band of white hairs, as in the male.

The palpus is pale, covered with white hairs, and having a fringe of black hairs at the extremity.

The legs are varied with rufous and black.

The abdomen is reddish on the base and anterior sides, the posterior sides and apex being whitish. Near the base, and close together, are two large, black spots, in the middle of each of which is a small white dot. The black spots narrow into bands, and curve outward, and then inward again, enlarging into two posterior spots which are joined in the middle by a short, transverse black band, which curves forward. The central region thus enclosed is whitish, having a darker spot in the middle. Posteriorly, and growing smaller toward the apex, there are three black spots, the anterior and largest of which lies within the curve of the short transverse band which unites the posterior spots. The spinnerets are reddish.

The epigynum presents a plate twice as wide as long, with a curved opening at the posterior edge.

Habitat, Wisconsin.

XXI. ATTUS MAXNII, new.

(Plate III., Figure 21.)

MALE. Length, 8 mm.; width of abdomen, 2.7 mm.; length of ceph.th., 4.1 mm.; width of ceph.th., 2.5 mm.

Legs.	Fem.	Pat.	Tib.	Metat.	Tar.	Total.
1.	2.1	1.5	2.1	1.2	.9	7.8
2.	2.	1.5	1.7	1.2	.8	7.2
3.	2.5	1.1	1.1	1.2	.8	5.9
4.	2.6	1.1	1.9	1.4	.9	7.9

The cephalothorax is low and flat; the thoracic part is longer than the cephalic; in color it is jet black, probably originally covered with mixed gray and tawny hairs. The lower narrow margin is black, and above this is a band of white hairs. Another white band, which is a continuation of the white color of the clypeus, extends around the anterior sides, midway between the dorsal eyes and the lower white band. The fovea is deep.

The quadrangle of the eyes is wider than long, and is not wider behind. The dorsal eye is as large as the lateral eye, and is on a line with it. The small median eye is equidistant from the dorsal and lateral eyes, and is on a line with their upper borders. The anterior eyes are surrounded with red hairs; they are in a slightly curved row, a straight line from the top of the middle eyes cutting the lateral eyes above the centre. The middle eyes are sub-touching, and are twice as large as the lateral eyes, which are more widely separated. The clypeus is one-fourth as wide as the middle eyes; it is vertical, and has red hairs just below the eyes, and long white hairs on its lower edge.

The palpus has a simple bulb, which does not project backward over the tibia.

The falces are vertical, and moderately long and stout. In color they are black, with a transverse fringe of long white hairs where they meet at the base.

The maxillæ and lip are blackish. The maxillæ are wider at the extremity, and rounded. The lip is half as long as the maxillæ, wide, and also rounded.

The sternum is narrower in front than behind; it is black, with long white hairs.

The relative length of the legs is 4, 1, 2, 3. The first pair is a little the stoutest. In color they are black, or dark brown, with patches of white and tawny hair. There are spines on the tibial and metatarsal joints of the four pairs; and on the femoral joints of the second, third and fourth. The spines on the metatarsi of the fourth extend to the base.

The abdomen is flattened. There is a central longitudinal band, black, and on either side of this a band of coarse gray hair.

The venter is blackish in the central line, and white on the sides. Habitat, Florida.

Observations. We have named this species for Mr. Chas. Mann, of Milwaukee, Wis.

•

XXII.   EPIBLEMUM PALMARUM, Hentz.

(Plate III, Figures 22 to 22a.)

MALE—Length, 5.5 mm.; width of abdomen, 1.1 mm.; length of ceph.th., 2.1 mm.; width of ceph.th., 1.8 mm.

The cephalothorax is not high, the cephalic part being much the shorter. There is a deep depression between the cephalic and thoracic parts. The color is bronze brown with short golden down, lighter in the eye region. The lower narrow margin is black. A wide white band passes around the clypeus and the upper sides just below the eyes; just behind the dorsal eye, on each side, this band curves upward over the thoracic part, and ends near the base. The eyes are surrounded with bright red hair.

The quadrangle of the eyes is wider than long, and not wider behind. The dorsal is as large as the lateral eye, and is on a line with it, the small median eye being on a line with their upper borders. The anterior row of eyes is straight. The middle eyes are touching; the lateral eyes almost touch the middle eyes, and are half as large. The clypeus is half as wide as the middle eyes; it is vertical, and covered with long, snowy white hair.

The palpus is relatively shorter and smaller than in E. scenicum. The femur is slender; the patella is twice as long as the tibia, and a little stouter. The tibia has a slender, curved, pointed apophysis, which is one-fourth as long as the tarsus. The tarsus is as long as the tibia and patella together. The bulb is not complicated.

The falces are long, but not so long, in proportion to the cephalothorax, as in E. scenicum. They are horizontal, flattened above, convex below. There are two teeth, not three, as in E. scenicum. The hook is as long as the rest of the falx.

The maxillæ are long, truncated at the top, projecting at the outer corner, and straight on the inner border. The lip is two-thirds as long as the maxillæ, and is pointed.

The sternum is long, rufous. The anterior thighs are separated by more than the width of the lip at the base.

The relative length of the legs is 1, 2, 3, 4. The legs of the first pair, which are much the longest and stoutest, are bent, and are dark reddish brown in color. The other legs are pale and weak. There are femoral, tibial and metatarsal spines on the four pairs; those on the femora of the first pair are small; those on the metatarsi of the fourth are found only at the extremity.

The abdomen is slender, and is of a bronze brown color, covered with golden down. There are white bands on the sides, from base to apex. These bands do not meet at the base, where there is a tuft of black hairs.

The venter is pale rufous, edged with white.

FEMALE—Length, 4.1 mm.; width of abdomen, 1.2 mm.; length of ceph.th., 1.6 mm.; width of ceph.th., 1.4 mm.

The cephalothorax is rufous covered with short white hairs. The lower narrow margin is black. There are two dark spots in the eye-region just back of the anterior middle eyes, and some long black hairs just outside of the lateral eyes. The dorsal and small median eyes are situated on small elevations.

The palpus is pale with three distinct dark rufous bars.

The falces are rather stout ; they are short and vertical, and dark reddish brown in color.

The maxillæ are long, and enlarged at the tip.

The abdomen is pale, covered with white hairs and having a tuft of white hairs at the base. There are some dark rufous markings, indistinctly visible through the hairs, which seem to consist of three transverse bands on the dorsum, and some oblique bands on the sides. When the hairs are rubbed off, the dorsal markings are seen to be a branching central longitudinal band, which is made up of many dots, spots and lines.

The venter is pale, with a reddish central band, and is covered with white hairs.

The epigynum presents a heart shaped plate, with an opening at the posterior edge, and two round elevations, one on each side.

Habitat, New Jersey, North Carolina, South Carolina, Florida.

Observations. Thorell is certainly right in saying that E. palmarum cannot belong to the same genus as E. faustum, of Hentz.

XXIII. SYNEMOSYNA FORMICA Hentz.

(Plate III, Figures 23 to 23a.)

MALE—Length, 3.9 mm.; width of abdomen, .7 mm.; length of ceph.th., 1.8 mm.; width of ceph.th., .8 mm.

Legs, 3, 2.2, 2.5, 3.2.

The cephalothorax is low ; the thoracic part is much longer than the cephalic, and has two constrictions, one just behind the dorsal eyes, forming a high ridge, and a second just before the base. In color it is brown, lighter on the upper surface, sometimes blackish on the sides, smooth, glabrous, with a few short white hairs in the eye-region.

The quadrangle of the eyes is wider than long, and a very little wider behind. The dorsal eye is slightly larger than the lateral, and is higher, a straight line from its lower border passing above the lateral eye. The small median eye is equidistant from the lateral and dorsal eyes, and is on a line with the lower border of the dorsal. The anterior row of eyes is slightly curved, a straight line from the top of the middle eyes cutting the lateral eyes above the centre. The middle are four times as large as the lateral eyes, and all four are touching.

The clypeus is one-fourth as wide as the middle eyes. It is vertical, and blackish, without hairs.

The palpus has a stout, blunt, tibial apophysis. The bulb is rounded. The tube is long ; it curves once around the bulb, and then passes to the extremity of the tarsus.

The falces are weak, short, vertical, black, with pale edges.

The maxillae are enlarged and truncated at the tip ; their color is white, with a black central spot. The lip is wider than long, less than half as long as the maxillae, truncated at the tip, black, edged with white.

The sternum is long. The anterior half is pale. The posterior half blackish. The anterior thighs are separated by the width of the lip at the base.

The relative length of the legs is 4, 1, 3, 2. They are all slender. The first leg is pale, with an internal and an external black line on the femur, patella, and tibia. The second leg is all pale. The third leg is pale excepting the femur, which is light rufous. The fourth leg has the femur rufous; the proximal end of the patella pale ; the distal end of the patella, and the proximal end of the tibia, blackish, shading into pale at the distal end of the tibia ; and the metatarsus and tarsus pale. There are a few spine-like hairs on the tibia and metatarsus of the first and second legs.

The abdomen has a deep constriction before the middle, posterior to which it is enlarged and rounded. The part before the constriction is pale rufous with a darker longitudinal band on the upper surface. Posterior to the constriction it is black, with a narrow pale band. The spinnerets are pale.

The venter is pale in front, black posteriorly.

FEMALE—Length, 5.4 mm.; width of abdomen, 1.8 mm.; length of ceph.th., 2.2 mm.; width of ceph.th., .7 mm.

Legs, 2.7, 2.2, 2.8, 3.5.

The abdomen has a slight constriction near the base, which is slender. Just behind the second and deeper constriction is a pale band, which is wider than in the male, occupying the anterior sides, and curving downward under the venter.

The epigynum presents a raised plate, with a single opening at the posterior edge. The tubes, which are visible through the integument, take somewhat the form of a figure 3.

Habitat, North Carolina, Alabama, Washington, D. C., Massachusetts, Pennsylvania, Illinois, Wisconsin, Iowa.

Observations. We find the falces short in the males, as did Mr. Emerton. Hentz is probably in error on this point. For a good drawing of the palpus of male, see plate 20, fig. 9, of Hentz's Arachnological Writings, edited by Burgess.

XXIV.   ATTUS CARDINALIS Hentz.

(Plate III, Figures 24 to 24a.)

MALE. Length, 10 mm.; width of abdomen, 3.5 mm.; length of ceph.th., 4.8 mm.; width of ceph.th., 4.3 mm.

Legs, 10.5, 8.5, 8.5, 11.2.

The cephalothorax is rather low and wide. The thoracic part is longer than the cephalic, and is separated from it by a distinct depression. The cephalic part is a little inclined; the thoracic, slopes abruptly to the base. The color is dark on the sides and base, while the upper surface is covered with short red hair, with longer black hairs scattered over it.

The quadrangle of the eyes is wider than long, and plainly wider behind. The dorsal eye is as large as the lateral eye, and is placed higher, a straight line from the lower margin of the dorsal cutting the upper margin of the lateral eye. The median eye, which is very small, lies in a line with the middle of the lateral eye, and is nearer the lateral than the dorsal eye. There is a small elevation just above and within the dorsal eye. The anterior eyes are almost touching, and are in a slightly curved row, a straight line from the summit of the middle eyes cutting the lateral eyes above the centre. The lateral are about half as large as the middle eyes. The clypeus is about one-fourth as wide as the middle eyes, vertical, black, with a few long reddish hairs.

The palpus is long, with many reddish hairs. The tibia is shorter than the patella, with an apophysis on the outer side. The bulb of the tarsus is long and simple.

The falces are stout and rather long; they are vertical, and vary in color from blue to green.

The maxillæ are rounded and dark colored; the lip, also dark colored and rounded, is half as long.

The sternum is oval, black. The anterior thighs are separated by the width of the lip at the base.

The relative length of the legs is 4, 1, 3, 2. The first pair is the stoutest, the second next. The femora are dark, the other joints brown or reddish. There are tibial and metatarsal spines on the four pairs, and femoral spines on the first and second.

The abdomen is uniform scarlet, or, in some specimens, brick red. There are four impressed dots on the dorsum.

The venter is dark.

FEMALE—Length, 10 mm.; width of abdomen, 4.9 mm.; length of ceph.th., 4 mm.; width of ceph.th., 3.1 mm.

Legs, 7.5, 6, 6.4, 8.1.

The abdomen has a whitish line around the base.

The venter has two longitudinal white lines running to a point in front of the spinnerets.

The epigynum presents a rounded plate, with two dark spots and a large posterior opening.

Habitat, North Carolina, Pennsylvania, New York, Wisconsin, Iowa.

Observations. Not uncommon in Wisconsin.

XXV. Attus trifunctatus Hentz.

(Plate III, Figure 25.)

MALE. Length, 11. mm.; width of abdomen, 4.3 mm.; length of ceph.th., 5.3 mm.;
width of ceph.th., 5.1 mm.

Legs, 1, 9.3, 9.1, 10.7.

The cephalothorax is high, the sides curving outward from the
dorsal eyes, which are placed upon the upper surface rather than upon
the sides of the caput. The thoracic part is longer than the cephalic.
In color it is black, covered with short black and gray hairs, and having some long black hairs on the sides near the small median eyes.
There are sometimes white bands on the sides.

The quadrangle of the eyes is wider than long, and plainly wider
behind. The dorsal eye is as large as the lateral, and is placed higher,
so that a straight line from its lower border passes above the lateral
eye. The small median eye is much nearer the lateral, and is so
placed that its lower border is on a line with the upper border of the
lateral, while its upper border is on a line with the lower border of the
dorsal eye. The anterior row of eyes is slightly curved, a straight
line from the top of the middle eyes cutting the lateral eyes above
the centre. They are all well separated, the lateral more widely from
the middle eyes than the middle eyes from each other. The middle
are more than twice as large as the lateral eyes. The clypeus is half
as wide as the middle eyes, and is vertical; its upper half is covered
with short, and its lower half with long white hairs.

The femur of the palpus is long and rufous, with a black band on
the inner side, three black spines, and some black and white hairs.
The patella is longer and stouter than the tibia. The bulb of the tarsus projects backward to the extremity of the patella.

The falces are stout, short, vertical, and somewhat inclined forward. Their color is bright iridescent green. The hooks are reddish
brown.

The maxilla, which is nearly as wide as, and parallel with the thigh
of the first leg, is nearly as long as the coxa and exinguinal joint together. It is obliquely truncated, and has a short projection from its
outer corner. The lip is half as long, rounded.

The sternum is oval, black. The anterior thighs are separated by
more than the width of the lip at the base.

The relative length of the legs is 1, 4, 2, 3. The first pair is the
stoutest, the second next. The femur and tibia of the first leg are
much stouter than the patella. The legs are black and hairy. On the
inner side of the patella of the first leg is a brush of white hairs.
Immature specimens have the legs barred with rufous. There are

5

spines on the femoral, tibial, and metatarsal joints of the four pairs, those on the metatarsi of the fourth extending to the base.

The abdomen is black, covered with thick, short, black hairs, and having some long white hairs. At the middle point is a large, more or less triangular white spot ; posterior to this are two smaller somewhat oblique white spots ; lower down and nearer the apex than these, but in a line with them, are two minute white dots. The spots are formed by scales, which through the microscope look like grains of rice. In some specimens there are two oblique white bands on each side, and a white band at the base. Some immature specimens have the spots orange colored instead of white.

The venter is black with two whitish bands which approach each other, but terminate near the apex without meeting.

FEMALE- Length, 15 mm.; width of abdomen, 5.6 mm.; length of ceph.th., 5.5 mm ; width of ceph.th., 4.4 mm.

Legs, 10.6, 9,9, 12.1.

The cephalothorax is somewhat iridiscent in the region of the eyes.

The palpus is long, rufous in color, with short white hairs at the joints, and otherwise covered with mixed black and white hairs. At the extremity of the tarsus is a thick brush of short dark hairs.

The maxillæ are rounded at the tip, instead of truncated, as in the male.

The relative length of the legs is 4, 1, 2, 3.

The epigynum presents a plate much longer than wide, with a large posterior opening.

Habitat, United States.

Observations. This species seems to be much larger in the Western than in the Eastern States. We have so far received no specimens from west of the Rocky Mountains. Attus audax of Hentz is probably a variety of tripunctatus. We have examined several specimens that had the markings of audax, and have found their epigynes to be like that of tripunctatus.

A period of from fourteen to fifteen days is required for the development of the eggs of this species.

For a good drawing of the male palpus of tripunctatus, by Mr. Emerton, see plate XX, figure 6, of Hentz's Spiders of the United States, edited by Edward Burgess, Boston Society of Natural History, 1875.

G. W. AND E. G. PECKHAM.   36

EXPLANATION OF PLATES.

PLATE I.

1. ATTUS PUTNAMII, dorsal markings of male, x 3; 1a, male palpus; 1b, anterior eyes and falces of male, x 6.
2. ATTUS ÆSTIVOLIS, dorsal markings of female, x 2; 2a, epigynum; 2b, male palpus; 2c, maxillæ and falces of male.
3. ATTUS SPLÆNDENS, dorsal markings of female, x 4; 3a, epigynum; 3b, male palpus.
4. ATTUS OCTO-PUNCTATUS, male palpus.
5. ATTUS HOYI, male palpus.
6. ATTUS FLAVUS, male palpus.
7. ATTUS RUSTICOLUS, epigynum.
8. ATTUS TIBIALIS, male palpus; 8a, epigynum.
9. ATTUS AGRESTIS, dorsal markings of female, x 3; 9a, epigynum.

PLATE II.

10. ATTUS ARIZONENSIS, dorsal markings of male, x 2; 10a, male palpus.
11. ATTUS MINIATUS, dorsal markings of female, x 2; 11a, epigynum.
12. ATTUS M'COOKII, epigynum.
13. ATTUS PERRORINUS, male palpus; 13a, third leg of male, from the posterior side.
14. ATTUS PRINCEPS, epigynum.
15. ATTUS QUADRILINEATUS, epigynum.
16. ATTUS PINUS, epigynum.
17. ATTUS JOHNSONII, epigynum; 17a, male palpus.
18. ATTUS FORMOSUS, abdomen of female, x 6.
19. ATTUS ALBO-IMMACULATUS, epigynum.

PLATE III.

19a. ATTUS ALBO-IMMACULATUS, front view of first leg of female, x 10.
20. ATTUS PALUSTRIS, dorsal markings of male, x 8; 20a, male palpus; 20b, abdomen of female, x 2; 20c, epigynum.
21. ATTUS MANII, male palpus.
22. EPIBLEMUM PALMARUM, male palpus; 22a, epigynum.
23. SYNEMOSYNA FORMICA, side of female, x 4; 23a, epigynum.
24. ATTUS CARDINALIS, epigynum; 24a, male palpus.
25. ATTUS TRIPUNCTATUS, epigynum.

NOTE.—The epigynes are all drawn as they appear under alcohol.

PLATE 1

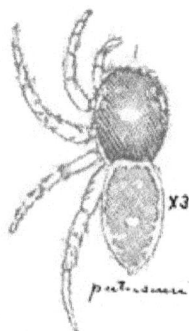

putnami

putnami

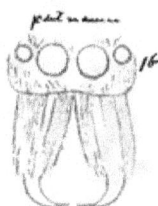

aestivalid

aestivales

aestivalis

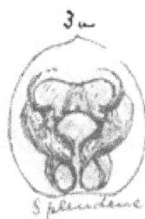

Splendens

Splendens

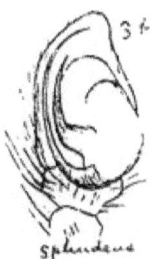

Splendens

sili-punctatus

kaye

flavus

tibialis

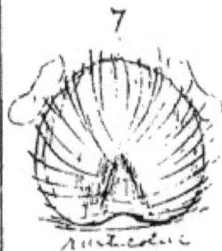

auticolus

tibialis

agrestis

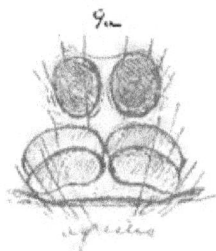

agrestis

PLATE II.

10

×2

arizonensis

10a

arizonensis

11a

miniatus

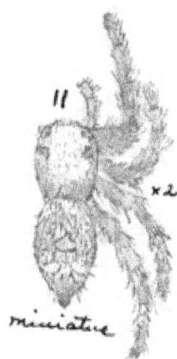

11

×2

miniatus

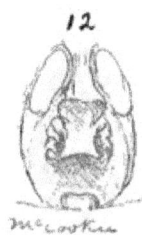

12

McCooki

13

peregrinus

13a

peregrinus

14

princeps

17

johnsoni

15

q. m. intimatus

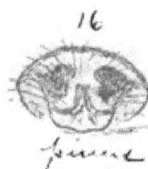

16

juven.

17a

johnsoni

18

×6

formicus

19

albo-immaculatus

RUDOLPH HAESSLER. From Nature.   CHAS F SASSE Photo Lith New York

PLATE III

20

palustris

20a

19a

×10

20b

×6

punctata

albo-maculatus

21

maccei

22

palmarum

22a

palmarum

23a

formica

23

formica ×4

20c

palustris

24

cordinalis

24a

cordinalis

25

bipunctatus

www.ingramcontent.com/pod-product-compliance
Lightning Source LLC
Chambersburg PA
CBHW022030190326
41519CB00010B/1656